John Jennings Moorman

White Sulphur Springs

With the Analysis of its Waters, the Diseases to which They are Applicable

John Jennings Moorman

White Sulphur Springs
With the Analysis of its Waters, the Diseases to which They are Applicable

ISBN/EAN: 9783337144272

Printed in Europe, USA, Canada, Australia, Japan

Cover: Foto ©berggeist007 / pixelio.de

More available books at **www.hansebooks.com**

White Sulphur Springs,

WITH THE

ANALYSIS OF ITS WATERS,

The Diseases to which they are Applicable,

AND SOME ACCOUNT OF

Society and its Amusements at the Springs,

BY

J. J. MOORMAN, M. D.,

Physician to the White Sulphur Springs : Author of Mineral Waters of the South and Southwest ; of Mineral Springs of North America, &c. &c.; late Professor of Medical Jurisprudence and Hygiene in the Washington University, Baltimore; Member of the Medico-Chirurgical Society of Maryland ; of the Baltimore Medical Association, &c., &c., &c.

BALTIMORE:
THE SUN BOOK AND JOB PRINTING OFFICE
1878

INDEX.

☞ *See Announcement at close of Pamphlet, and* CARD *of Proprietors as to the Non-use of the waters,* except to guests on the Grounds.

WHITE SULPHUR SPRINGS,

Greenbrier County, West Virginia.

PAMPHLET for general circulation, adapted for a HANDBOOK, to guide the distant stranger as to the *location* and *extent of accommodations* of the WHITE SULPHUR SPRINGS—to point out the different *routes* by which they may be reached—and, at the same time, indicating in a concise manner the *various diseases* for which their waters have been advantageously used, has long been a desideratum with the Spring-going public.

In attempting to supply this want, by the issue of this pamphlet, I shall not insert general certificates in proof of the value and medicinal adaptations of the waters to the various diseases for which they have been so long successfully used. The publication of such certificates, while they might be serviceable in some cases, would, nevertheless, be liable to mislead from the want of proper and scientific discrimination as to the precise nature of the cases given.

Mineral waters, to establish and perpetuate a valuable reputation, must be carefully kept *within the clear boundary of their power over disease*, and *within their true adaptation as curative agents*. To be efficiently employed, whatever be the *name* of the *disease* for which they are used, the *state of the system* at the time, and a proper administration, so as to secure desired effects, are important points that cannot be safely ignored.

There are biasing partialities and prejudices in the whole *certificate system* that are sure to intrude in despite of every effort to keep them out, and hence it is that conclusions in such cases are apt to be too sweepingly made to be realized by those who rely upon them. I have had abundant reason to know that great injustice is often done to suffering humanity, however unintentionally—and ultimately, too, to mineral waters, by having them placed, through vague and extravagant certificates, upon the common platform with patent medicines. In this way hopes are often created in the minds of invalids that are destined to sad disappointment; while the failure of the waters to accomplish all that had been injudiciously promised for them causes their reputation unduly to suffer in public estimation.

Mineral waters possess great and valuable powers, and are in many cases superior to the medicines of the apothecary's shop;

and, when used under proper and judicious discrimination, are well qualified to assume a place in the great medical mind of the world, and, like well defined articles of the *Materia Medica*, stand prominently forth as most valuable resources of the healing art.

These views are sanctioned by a sufficient amount of truth and importance to influence me against the common practice of publishing certificates of cases of diseases, unless such cases had been carefully *diagnosed* by a party competent to such duty, and so clearly described as to give them a fair claim to an intelligent public reliance.   Upon this branch of the subject, therefore, I propose to rely upon the general results of public opinion, formed from the use of the water for *nearly a century*, and from my own professional experience in their administration for *forty years*, in cases the *precise pathology and nature of which were carefully investigated in connection with their use in each case*, enabling me, I conceive, to determine their power and applicabilities with the certainty that physicians determine the peculiar action of any article of the drug shop with which they are most familiar.

LOCALITY OF THE SPRINGS.

The *White Sulphur Springs* are situated on Howard's Creek, in *Greenbrier County, West Virginia*, and upon the western slope of the great Apalachian chain of mountains which separates the waters that flow into Chesapeake Bay from those that run into the Gulf of Mexico.

The situation of the Spring is elevated and beautifully picturesque, surrounded by mountains on every side. *Kates Mountain* is in full view, and about two miles to the south; to the west, and distant about one mile, are the *Greenbrier Mountains*, while the towering Alleghany, in its magnificent proportions, is found five miles to the north and east.

The Spring is in the midst of the celebrated "Spring Region," having the "*Hot*," "*Warm*" and "*Healing Springs*" from thirty to thirty-five miles to the north; the "*Sweet*" and the "*Sweet Chalybeate*," sixteen miles to the east; the "*Salt*" and the "*Red Sulphur*," the one twenty-four, the other forty-one miles to the south.

Its latitude is about 37½° north, and its longitude 3½° west from Washington.   Its elevation above tide water is 2,000 feet. The temperature of its waters is 62° Fah., from which they do not vary during the heat of summer or the cold of winter.

The Spring yields more than *thirty gallons* a minute; and it is a remarkable fact that this quantity is not perceptibly varied during the longest spells of wet or dry weather.   The quantity and temperature of the Spring being uniform under all circumstances, give a confidence, which experience has verified, of its uniform strength and efficiency.

It is surrounded by mountain and intervale scenery of great beauty, and blessed with a most delightful summer and fall climate. Independently of the benefit to be derived from the waters, a better situation for a residence of invalids and delicate persons during the summer and fall months can scarcely be imagined. They have here the advantage of a most salubrious and invigorating air and the most agreeable temperature—cool at morning and evening, and at no time oppressively warm. The thermometer ranges here during the summer between 60° and 75°, and rarely attains a greater height than 85° at any time of the day, while the atmosphere is so elastic and invigorating as to enable invalids to take exercise in the open air without inconvenience or fatigue.

ROUTES TO THE SPRINGS.

The *White Sulphur* is immediately on the Chesapeake and Ohio Railroad, 100 miles west of *Staunton*. This road is now completed to *Huntingdon*, on the Ohio river, thus rendering the Springs approachable by rail both from the East and West.

☞ Travelers from the *North* or *East*, by rail, must necessarily make the city of *Staunton* a point in their line of travel.

☞ The *route* to the Springs from *Washington* is by way of the *Orange and Alexandria* Railroad to *Gordonsville ;* thence on to the *Chesapeake and Ohio* Road, by the way of *Staunton*, or by *Harper's Ferry*, and up the Shenandoah Valley to *Staunton*.

☞ Persons coming from the *West* or *Southwest*, may travel either by way of *Washington* or *via Cincinnati* to *Huntingdon*, the terminus of the Chesapeake and Ohio Road, and thence on this road about 160 miles to the Springs.

The *route* from Cincinnati by way of Huntingdon is several hundred miles shorter than the old route by Washington.

☞ Those who wish to reach the Springs from the *South* have a continuous chain of railroad either by way of *Richmond* or *Knoxville*, Tennessee.

If the Knoxville route is taken, the traveler proceeds by way of *Lynchburg* to *Charlottesville*. At the latter place he takes the cars of the Chesapeake and Ohio Road for the *White Sulphur*, 140 miles distant.

☞ The *time* from Washington to *White Sulphur* is about fifteen hours.

EXTENT AND CHARACTER OF ACCOMMODATIONS.

In the spring of 1857 this property was purchased by a company of gentlemen, residing principally in Virginia, who, in virtue of an act of the Legislature, associated themselves in a joint stock company, under the name of the *"White Sulphur Spring Company."*

In conformity with the public demand for a large extension of accommodations, the Company immediately entered upon an extensive system of improvement, designed alike to increase the capacity of the property for the accommodation of visitors, and at the same time to beautify and adorn the grounds. To these ends they have erected the largest building in the Southern country— its dimensions being 400 feet in length by a corresponding width, and covering more than an acre of ground. This building is appropriated for *receiving rooms, dining room, ball room, parlors, lodging rooms, etc.* The parlor is one of the most elegant and spacious saloons in America, being half as large again as the celebrated "East Room" in Washington. The dining room is one of the largest in the world, being upwards of 300 feet long by a corresponding width, and conveniently seating 1,200 persons.

The Company has also built a large number of handsome *cottages* for families. In several respects the grounds have been greatly improved, particularly by the construction of broad serpentine walks in various directions through the lawns, and by widening and extending the romantic and popular *Stroll*, known as the "Lover's Walk." With these improvements, together with a new and capacious *bathing establishment*, and the removal of many of the old buildings to new locations, by which the lawns are enlarged and adorned, the property, alike in capacity, in convenience, and in the elegance of its arrangements, exhibits a new and greatly improved appearance.

ANALYSIS OF THE WATERS.

The White Sulphur was analyzed in the winter of 1842, by Prof. Hayes, of Boston, from a few bottles of the water sent to his laboratory the preceding fall. From his report, 50,000 grains (about seven pints) of this water contains in solution 3,633 water grain measure of gaseous matter, or about 1.14 of its volume, consisting of

```
Nitrogen Gas........................................ 1 013
Oxygen Gas.......................................... .108
Carbonic Acid....................................... 2.244
Hydro-Sulph. Acid * ..............................; .068
```

One gallon, or two hundred and thirty-seven cubic inches of the water, contains 19 739-1000 cubic inches of gas, having the proportion of

```
Nitrogen Gas........................................ 4.680
Oxygen Gas.......................................... .498
Carbonic Acid.......................................11.290
Hydro Sulph. Acid................................... .271
```

* It must be borne in mind that this water was examined by Prof. Hayes several months after its removal from the Spring, and consequently after it had parted with a large portion of its free hydro-sulph. acid gas.

Fifty thousand grains of this water contain 115 735-1000 grains of saline matter, consisting of

Sulphate of Lime... 67.1GS
Sulphate of Magnesia...... ···............... .... 30.3G4
Chloride of Magnesium .... .859
Carbonate of Lime... G.060
Organic Matter (dried at 212°).............................. 3.740
Carbonic Acid.. 4.584
Silicates (Silica 1 34, Potash 18, Soda 66, Magnesia and a trace
   of Oxide of Iron)..................... .... 2.960

Professor Hayes remarks that the organic matter of the water, in its physical and chemical character, differs essentially from the organic matters of some thermal waters; in contact with earthly sulphates, at a moderate temperature, it produces hydro-sulphuric acid, *and to this source that acid contained in the water may be traced.* He adds:

"The medicinal properties of the water is probably due to the action of this organic substance. The hydro-sulphuric acid, resulting from its natural action, is one of the most active substances within the reach of physicians, *and there are chemical reasons for supposing that after the water has reached the stomach similar changes, accompanied by the products of hydro-sulph. acid, take place.*" *

Professor William B. Rogers also analyzed this water, with the following results:

Solid matter, procured from 100 cubic inches, dried at 212° Fah., consisting of 65-54 grains.

Sulphate of Lime................................... 31.680 grains.
Sulphate of Magnesia............................... 8.241 "
Sulphate of Soda.. .... 4 050 "
Carbonate of Lime.................................. 1.530 "
Carbonate of Magnesia.............................. 0 506 "
Chloride of Magnesium.............................. 0.071 "
Chloride of Calcium................................ 0.010 "
Chloride of Sodium...... .... 0.226 "
Proto-Sulphate of Iron............................. 0.069 "
Sulphate of Alumniæ................................ 0.012 "
Earthy Phosphates—a trace
Azotized Organic Matter, blended with a large propor-
   tion of Sulphur, about.. ·...................... 0.005 "
Iodine, combined with Sodium or Magnesium.

Volume of each of the gases, in a free state, estimated in 100 cubic inches: †

Sulphuretted Hydrogen...................................0.66
Nitrogen...1.88
Oxygen 0.19
Carbonate Acid..3.67

* See Chapter III, on the "Relative virtues of the saline and gaseous contents of the White Sulphur Water," in the "*Mineral Springs of North America,*" by the Author.
† 100 cubic inches amount to about three and a half pints.

MEDICAL CHARACTER OF THE WATER.

The destinctive medical influences of this water upon the system are *Cathartic, Diuretic, Sudorific,* and *Alterative.*

Some *cathartic* and *diuretic* effect, as well as a distinct determination to the *skin by sweating,* is easily induced under its use, in the great majority who drink it. But the most decidedly controlling effect of the water over diseased action, and that which, more than every other, gives its highest and most valuable character as a remedy, is its ALTERATIVE POWER, or that peculiar action by which it effects salutary *changes* or *alterations* in the blood, in the various secretions, and upon the various tissues of the body.

The certain effects of the water in *stimulating glandular secretions, dissolving chronic inflammations, overcoming obstructions,* and throwing off offensive *debris* from the diseased system, leave no doubt of its distinctive and active *alterative powers.* Indeed, no article of the Materia Medica has more decided alterative effects.

I desire, especially, to call the attention of physicians, and the intelligent public generally, to this *distinctive and remarkable quality of the water.* In this, more than anything else, it differs from other mineral waters. Many other waters are found to possess valuable alterative power, and with an equal or greater cathartic or diuretic action, but none have yet been shown to be so *certainly, promptly* and *powerfully alterative* upon the human system.

Some of my unprofessional readers may desire to know the precise meaning that is attached to the term ALTERATIVE, in a medical sense. This term simply means to *alter* or *change ;* that is, to alter or change the chemical composition of the blood, the secretions of the glands and the various secretory organs and surfaces, the removal of obstructions from the glands or minute vessels which occur in congestions, irritations and inflammations; thus restoring the blood and the general organism to their natural condition, and to the performance of their natural functions.

I claim that the water has these effects by being *absorbed,* or, in other words, *entering into the great circuit of the circulation,* and thus exercising the specific or peculiar action of its constituents in promoting the various secretory and excretory processes, and thereby restoring the diseased system to a physiological condition.

Such effects and changes, wrought in the sick body, are obviously an *alteration,* and the remedy that produces them is an alterative.

This is but a part of a medicinal alterative, but it conveys a sufficient idea of its nature.

The water is also remarkable for its power in *reducing the force and frequency of the pulse,* when unduly excited. Its influence in this respect should be regarded, not as a *direct sedative effect* of the agent, but as the result of its potency in abating general excitement, resolving inflammations, and removing obstructions thus bringing back the system to its normal condition.

Experience in the administration of these waters has abundantly established the fact of their *direct* and *positive* influence in controlling and eradicating many diseases.   Their effects, when properly used, are to revive the languishing circulation, to give a new direction to the vital energies, re-establish the perspiratory action of the skin, bring back to their physiological type the vitiated or suppressed secretions, provoke salutary evacuations, either by urine or stool, or by transpiration; they bring about in the animal economy an internal transmutation, *a profound change.*   Entering the circulation, they course through the system, and apply the medicinal materials which they hold in solution, in the most minute form of subdivision that can be conceived of, to the diseased surfaces and tissues; they reach and search the most minute ramifications of the capillaries, and remove the morbid condition of these vessels which are so commonly the primary seats of the disease.

It is thus that they relieve chronic disorder action, and impart natural energy and elasticity to vessels that have been distended either by inflammation or congestion, while they communicate an energy to the muscular fibre and to the animal tissues generally which is not witnessed from the administration of ordinary remedies. It is thus that they produce the *alterative effect,* the *"profound change"* upon the system, of which I have been speaking.

It may be well to remark, that all mineral waters, to a greater or less degree, are stimulants, and consequently are inapplicable to the treatment of *acute* or *highly inflammatory* diseases. This is especially true of the White Sulphur, particularly when drunk fresh at the Spring and abounding in its stimulating gas.   It is true that when its volatile gas has flown off, it becomes *far less stimulating,* and may be used with safety and success in cases to which in its *perfectly fresh state* it would be unadapted.   But even in its least stimulating form, it is inadmissable for excited or febrile conditions of the system, and especially in high inflammatory action, at least until the violence of such action has been subdued by other agents.

But while these waters have excellent *adaptations* for the cure of many diseases, they are *unadapted* also to the treatment of others. It would be irrational to suppose that any medical agent capable of effecting so much good as is this water, when properly directed, should be incapable of doing harm when improperly used. In many cases in which it has a happy adaptation, it fails of its good effect from being *improperly* taken, while in a few cases its effects would be injurious from its very nature, however it might be employed. Some of the latter class of cases, very *prominent* in their importance, will be noticed under the distinct heading of *"Diseases in which the water should not be used."*

POPULAR ERRORS IN THE USE OF MINERAL WATERS.

Before entering upon the consideration of the diseases for which the White Sulphur Water may be beneficially employed, I desire for the benefit of invalids who may visit these, or other Mineral Waters, to make a few remarks with the view of correcting some, as least, of the *popular errors* that exist in connection with the use of such waters. And I trust that a life devoted to the investiga· tion of this general subject will relieve me from any appearance or egotism in doing so.

I regret that the limited space, within which I desire to confine this publication, will not allow me to say as much upon the subject as I could wish to do.

The first of the popular errors to which I call attention has reference to the *time invalids should use mineral waters.*

There is an opinion in the minds of not a few, that such waters should not be drunk longer than a *given number* of days, and, that after such time, they are prejudicial rather than beneficial. *This is an error.* There is *no specific time* during which invalids, speaking generally, should use this or any other mineral water. The time during which all such waters should be used depends upon the *nature of the case, the manner in which they are used*, the *susceptibilities of the case*, and their *effects* upon the organism and the disease.

There are periods, but *uncertain periods* in the use of the *White Sulphur Water*, when it ought to be suspended, or discontinued; but such periods can only be judged of by the *effects*, and not from any number of days during which it may have been drunk.

If the water be not adapted to the disease, and to the existing state of the system, of course it ought not to be used at all; but if it be adapted, it ought to be persisted in, until it produces its *alterative* or proper curative effects.

It happens in many cases, that some appropriate management is essential to prevent the water from having vagrant and undesirable operations, and to hasten its speedy and good effects; but it would be in vain to expect its use to result in a cure, until it has been properly employed for a *sufficient length of time* (and this irrespective of the number of days) to produce the *desired effect*.

From *two to eight weeks* is the range of time within which it may be made to produce all its good effects, or bring the system into such a condition as insures a return to health. But in less than *two weeks*, however skillfully directed, it need not be expected that it will be productive of its full sanitary influences.

The *second popular error* is that of hastily *changing from spring to spring*, without staying sufficiently long at any one to produce lasting or permanent good impressions upon the disease.

A restless disposition often causes invalids to fly from one spring to another, in the vain hope of greater good, when very probably the time they fruitlessly spent at several different springs would have been sufficient to cure them at any one of them, that might be even tolerably adapted to their condition.

This criticism does not apply to mere pleasure-seekers. They may properly go from spring to spring, and spend their time just where they are the happiest.

But it is not so with invalids who *have something for* the waters to do. They should wisely select the waters best adapted to their cases, and use them properly and perseveringly, until their unadaptedness is shown, or until they have produced such *effects* as the nature of the case demands. This being done, they can then with propriety resort to such other waters, or baths, as may be best adapted to their new condition.

The *popular errors* manifested in the *hap-hazard* and *experimental* methods of using mineral waters, are too numerous to be particularly considered in the space allotted to this subject, but they are too important to be entirely overlooked.

Potent mineral waters, that have been extensively used for many years, have, it is reasonable to suppose, established with those who have experienced, or long observed their effects, definite, and well defined laws of operation upon the human system; or, in other words, have established certain *hygienic* and *therapeutic* laws, by the observance of which they may be taken understandingly, safely, and in such manner to secure their sanative effects without incurring unnecessary delay or risk, from improper administration.

Nevertheless, many *serious invalids*, and those too who are desirous of speedy relief, will, with the utmost uncertainty of the correctness of such a course, enter upon the use of such waters entirely in an *experimental way*, and with as much disregard of the known laws of their administration, as if no such laws existed.

Such *experimentalists*, by *inefficient* or *untimely* dosing, or far more common, by *overdosing*, sometimes, by using the remedy when they are under a temporary excitement, or other influences that prevent its good effect (and which by a little precaution might speedily be removed), or by using it too *fresh* when it ought to be used *stale*, allow themselves either to be positively injured, or at least deprived of the benefits that might have resulted from its proper administration.

Such *tentative drinkers* may now and then adopt the best course that could have been pursued, and all will go well; but in no few instances it happens that grave mistakes are made, and real injury effected. But if no absolute injury shall have been done, it very commonly results that such experimentalists waste much time without deriving compensating advantages; and often, after a week or two of profitless experimenting, perceive the necessity of ceasing such a course, and begin the use of the agent *de novo* as it were, and as if they had never before taken it at all, and without having derived any advantage from the week or two, it may be, that they had been improperly using it.

Invalids intending to use mineral waters for the treatment of their diseases should first satisfactorily ascertain what particular water is best calculated to cure their maladies, and before entering upon its use, acquaint themselves with the proper way of using it and with the general management and precautions necessary to be observed while under its use, not only in reference to the *quantity, and times,* of drinking, but also in reference to *Baths,* the manner and periods of taking them, together with a *proper general rule of living* as relates to diet, *exercise* and *exposures,* while they are under agencies, and in a climate to which they have not been accustomed. Thus properly informed, and prudently conforming to judicious instructions, they may reasonably hope to secure all the advantages that can be derived from such agents.

The strength of the natural appetite for the good things of this world, and the *dicta* of fashion, but too commonly over-ride the most judicious medical advice at a fashionable watering place, in reference to *diet* and *dress.* Still I cannot conscientiously withhold this well-meant warning to the serious invalid at such places, to *eat prudently,* irrespective of the inducement which an active appetite may offer, and to *clothe warmly* and avoid *night exposure,* whatever fashion may demand or temptation crave.

BEST PERIOD OF THE YEAR FOR INVALID VISITATION.

I am often asked by correspondents and others, as to the best *time* or *period of the season for invalid visitation* to the Springs. As this is a matter of no little importance to the invalid, I remark in reference to the White sulphur, that from the *15th of May to the middle of July* is preferable to an *earlier* or *later period* of the season. There are substantial reasons why *invalids* should make their visits within the range of the time mentioned, and that they should prefer *an early* rather than a late period of this range of time.

1st—Because during this period we have the most delightful weather of the season; neither too warm nor too cool for exercise in the open air.

2d—Because the crowd of mere pleasure-seekers has not set in up to this period; the place is less crowded, and all the facilities and comforts of a quiet home are more easily and certainly obtained.

3d—In the early period of the summer solstice, just after the cold and inclement weather of winter and early spring, and before the sufferer has become enervated by the heat of the summer *chronic disease* more readily yields to the alterative influence of the waters, and, consequently, the invalid is more certainly and speedily placed under their curative powers; and

4th—Because invalids, whose maladies have been essentially modified or cured in the early part of the summer, have a longer period of favorable weather, either here or elsewhere, in which to perpetuate and confirm their amendment and final cure, than those who might receive influences equally beneficial, but obtained at a later period of the summer.

I might allude to other advantages enjoyed by the invalid who makes his visit to Mineral Waters early in the season; but let it suffice to remark, that my long observation as Medical Director of these waters has abundantly satisfied me of the decided advantage that attaches to early rather than late visitation by those who are seeking to secure the largest amount of benefit from their use. Hence, I earnestly suggest to *invalids* who design visiting these waters in the course of the season, not to postpone their visit to a late period of the season, and to *choose an early rather than a late* period of the time I have designated as preferable. Many invalids will derive as much advantage from three weeks' use of the water in June, as they will from four in September.

But while the summer, and especially the early portion of it, is the preferable time for using the waters, they may nevertheless be drunk to advantage during the cold weather of the late *fall*, *winter* or *early* spring, provided that those who use them are properly protected by clothing suitable for the season, and by warm and comfortable lodging.

DISEASES TO WHICH THE WHITE SULPHUR WATER IS APPLICABLE.

For want of space I can only give a mere synopsis of the diseases for the cure of which the White Sulphur Waters have been long and successfully employed, referring my readers who desire more particular information upon this subject, as well as for the *proper method of using the waters in the various diseases, to my volume on "The Mineral Springs of North America."*

DYSPEPSIA.

This very common and annoying disease, the especial scourge of the sedentary and the thoughtful, whether existing under the form of irritation of the mucous surface of the stomach, vitiation of the gastric juice, or under the somewhat anomalous characteristic of *Gastralgia*, is treated with much success by a proper course of the White Sulphur Water.

The apprehensive and dejected spirit that finds no comfort in the present, and forebodes only evil in the future; the hesitating will that matures no purpose, and desponds even in success; the emaciation of frame and haggardness of visage; the ever-present endurance, and all the imaginary and real ills that torture the hapless dyspeptic, are often made to yield to alterative and invigorating influences that a few weeks' judicious use of the waters have established.

Administered alone, *in every form* of this disease (for under the name of *Dyspepsia* we have several *forms of Stomach disease* essentially differing from each other, and requiring different modes of treatment) its curative powers may not always be so marked; but in several varieties of the disease, and those indeed which we most often witness, it deserves the very highest praise that can be conferred upon any remedy. In cases of this disease in which the *Liver* is implicated, occasioning slow or unhealthy biliary secretions, a state of things that often exists, the water may be used with special advantage. To effect *permanent* or lasting cures in Dyspepsia the waters should always be pressed to their complete *alterative* effects upon the system.

CHRONIC IRRITATION OF THE MUCOUS MEM-BRANE OF THE STOMACH AND BOWELS.

The largest class of invalids that resort to our mineral fountains for relief are those afflicted with *abdominal irritations*, and especially with *Irritations of the Mucous Coat of the Stomach and Bowels.*

These irritations are occasionally so marked by a super-added nervous mobility as to conceal their true character from the sufferer, and sometimes from his medical adviser. The disease is far more common in late than in former years. The number of cases at the White Sulphur has been, I am sure, more than triplicated within the last few years. It may be induced by any of the numerous causes whose tendency is to derange the digestive, assimilative and nervous functions; and is often connected with some indigestion, irregular or costive bowels, with restlessness and unhappy forebodings of impending evils. I have much confidence in the waters in such cases when prudently and cautiously used, aided, if necessary, by proper adjunctive means, and pressed to their full *alterative effects.**

---

* For a more full account of this disease, its symptons, and treatment, *see volume* on the "*Mineral Springs of the South and Southwest,*" by the *Author.*

LIVER DISEASES.

Chronic disease of the liver, in some form or other, is a very common disease of our country, especially in the warm latitudes and miasmatic districts. Very many affected with this complaint have annually visited the White Sulphur for the last fifty or sixty years. In no class of cases have the effects of the waters been more fully and satisfactorily tested than in *chronic derangements of the liver.*

The *modus operandi* of sulphur water upon the liver is dissimilar to that of mercury, and yet the effects of the two agents are strikingly analogous. The potent and controlling influence of the water over the secretory function of the liver must be regarded as a specific quality of the agent, and as constituting an important therapeutic feature in the value of the article for diseases of this organ. Its influence upon the liver is gradually but surely to unload it when engorged and to stimulate it to a healthy performance of its functions when torpid.

The control which this water may be made to exercise over the liver, in correcting and restoring its energies, is as often astonishing as it is gratifying—establishing a copious flow of healthy bile, and a consequent activity of the bowels, imparting a vigor to the whole digestive and assimilative functions, and, consequently, energy and strength to the body, and life and elasticity to the spirits.

For many years I have kept a *"Case-book"* at the White Sulphur, and have carefully noted the influences of the water upon such cases as have been submitted to my management. Among the number, are many hundred cases of chronic affections of the liver, embracing diseases of *simple excitement, chronic inflammation, engorgement,* and obstructions of the biliary ducts, etc. These cases were treated either with the White Sulphur alone, or aided by some appropriate adjunctive remedy; and, in looking at the results, I must be permitted to express a doubt whether a larger *relative* amount of amendments and cures has ever been effected by the usual remedies of the medical shop. This I know is high eulogy of the water in such diseases. It is considerately made, and is no higher than its merits justify.

When *Schirrosity* of the liver is suspected, the water, if used at all, should be used under the guard of a well-informed medical judgment; for in actual Schirrosity, if it be pressed beyond its primary effects upon the stomach and bowels, it is *very decidedly* injurious. I have known several cases in which death was hastened by disregarding this caution.

For a more full account of the influences of the water in Liver diseases, the reader is referred to the author's work on the *"Mineral Springs of North America.*

JAUNDICE.

This is a form of liver disease in which obstructions prevent the free egress of the bile from the gall-bladder along its natural channels, and hence occasions its absorption into the general circulation.

In cases of jaundice, in which the obstructing cause is inspissated bile, or very small *calculi*, or when occasioned by inflammation or spasm of the gall-ducts themselves, the White Sulphur Water, as might be expected from its influence over the liver, is used with the happiest results.

Indeed, the individuals affected with incipient or confirmed jaundice, and whose livers are free from scirrhus, cannot place too much confidence in the use of the White Sulphur Water and *baths*, with the occasional aid of mild adjunctive means to aid in its speedy action upon the liver and skin. Thus judiciously employed, and for a sufficient length of time, it invariably proves successful, either in curing the case or in bringing the system into the condition under which a cure speedily results.

CHRONIC DIARRHŒA.

In *Chronic Diarrhœa*, especially where the mucous coat of the bowels is principally implicated, and, still more, where the case is complicated with derangement of the stomach and liver, the Water is often employed with very gratifying effects.

While the Water, properly taken, is a most invaluable remedy in *Chronic Mucous Diarrhœa*, in no other disease are prudence and caution more eminently demanded in its administration, and especially for the first few days of using it. When prudently and cautiously prescribed in such cases, it is not only a perfectly safe remedy, but also eminently curative in its effects. Many of the most satisfactory results that I have ever accomplished by the prescription of the White Sulphur Water have been in cases of *Chronic Mucous Diarrhœa*.

SEROUS DIARRHŒA, of *chronic character*, requires still greater caution in the early use of the Water than the mucous form to which I have been referring; and while the Waters, when carefully introduced, constitute a valuable remedy in such cases, they will, if too largely taken, aggravate the worst symptoms of the disease.*

COSTIVENESS.

Habitual costiveness is a state of the system in which the Water has been extensively employed—sometimes successfully, sometimes not. When the case depends upon depraved or deficient

---

* See the details of several interesting cases in the "Mineral Waters of the United States and Canada," by the Author.

biliary secretions, much reliance may be placed upon the efficiency of this remedy, if it be carried to the extent of fully *altering* the system.

Costiveness, dependent upon *inertia* or loss of tone of the coats of the bowels, is among the most difficult of mere functional derangements to relieve. The persistent use of *alterative doses* of the Water will, however, sometimes effect it. But, most unwisely, many persons in this condition defeat their chance of a cure by the very improper practice of using *common salt* in the Water to render it purgative. Such a practice may answer a present purpose, but it does much to defeat the ALTERATIVE effects of the Water, which is its great power in such cases.

HAY FEVER.

This disorder, periodical in its attacks, and in its symptoms much resembling an ordinary catarrh, is more or less common to all latitudes. Without being dangerous in its consequences, it is annually annoying to many persons, and especially about the period of the fall equinox. Great mountain altitudes as a summer and fall residence, with tonics as medicine, are most relied upon for modification or cure.

The elevation of the White Sulphur, 2,000 feet above the sea level, with immediate surrounding mountains of 3,500 feet, together with an atmosphere of great purity and elasticity, and the alterative and invigorating effects of the Water, very happily adapts it as a place of summer and fall residence for those afflicted with Hay Fever. I have rarely, if ever, seen a case of this disease here that was not benefited; in some cases entirely relieved for the season, in others greatly modified. In all cases I think more or less benefit has been derived.

PILES.

The use of mild laxatives in *hæmorrhoids* has long been a favorite practice for their relief. The beneficial effects of the Water in this disease is probably to some extent due to its laxative power, but still more, I apprehend, to its *alterative effect* upon the liver, through which the hæmorrhoidal vessels are favorably impressed.

DISEASES OF THE URINARY ORGANS.

The White Sulphur Water is used with very good effects in *Gravel;* indeed, they almost invariably palliate such cases, and frequently, in their early stages, entirely cure them.

Incipient calculus affections are relieved by the Water pretty much in proportion as it corrects the digestive and assimilating functions, improves the blood, and brings the general economy into a natural type, preparing the kidneys to resist foreign encroachments upon their functions, and to elaborate, from healthy blood, proper and healthy secretions.* Where the affection depends upon *acid* predominance in the fluids, the Water never fails to palliate, and often cures the case. Whether or not this Water should be preferred to other remedies, in *calculus* affections, depends upon the *diathesis* that prevails in the system, and hence the urine should always be carefully analyzed, that we may not act in the dark in such cases.

Chronic Inflammation of the Kidneys, as well as similar affections of the bladder and urethra, are often successfully treated by a judicious use of the Waters. I have treated numerous cases of catarrh of the bladder successfully by a proper use of the Water and other appropriate remedies in connection with it, always regarding the Water, however, as the leading remedy in the case.

Diabetes is a form of disease in which the Waters have occasionally been used with excellent effect.

Spermatorrhœa, often painfully implicating the nervous system, and producing extreme debility, not only of the sexual organs, but also of the general system, is often greatly benefited at these Springs. This disease is generally found complicated with a condition of the skin and glandular organs, and not unfrequently of the mucous surfaces, that eminently require the aid of *alterative remedies*. In all such complications the Waters are found very valuable as a primary means, preceding and preparing the system for the use of more decided tonic remedies.

FEMALE DISEASES.

In *Female diseases*, in their various chronic forms of *amenorrhœa*, or suppressed menstruation, *dysmenorrhœa*, or painful menstruation, *chlorosis* and *leucorrhœa*, the waters of the White Sulphur have been much employed. When the cases have been judiciously discriminated, and were free from the combinations and states of the system that contra-indicate the use of the Waters, they have been employed with beneficial results.

CHRONIC AFFECTIONS OF THE BRAIN.

It is only since I inaugurated the custom of using the Water in its *ungaseous form* (thirty-eight years ago) that it has been taken successfully, or even tolerated by the system, in chronic inflammation of the brain. I need, therefore, scarcely apprise my readers that it is only in its strictly *ungaseous form* that it should be used in such cases, and then in a careful and guarded manner. Thus prescribed, I have in several instances found it beneficial.

---

* See "Mineral Waters of the United States and Canada," by the Author

NERVOUS DISEASES.

Neuralgia, in some form or other, has become a very common disease in every part of our country, and the number that visit the White Sulphur suffering with this *protean* and painful malady is very considerable. Sometimes this disease exists as a primary or independent affection, but far more frequently as a *consequence* of visceral or organic derangements. Where such is found to be the case, the White Sulphur Waters are used with the very best results. As an *alterative*, to prepare the neuralgic for receiving the more tonic waters to advantage, it deserves the largest confidence by those afflicted with this annoying malady.

PARALYSIS.

The number of paralytics that resort to the White Sulphur is large, and their success in the use of the Waters various. Cases resulting from dyspeptic depravities are oftener benefited than those that have resulted from other causes. In almost every case, however, some benefit to the general health takes place, and sometimes an abatement of the paralysis itself.

BREAST COMPLAINTS.

In *tubercular consumption*, whether the tubercules be incipient or fully developed, the White Sulphur Water should not be used. Its effects in such cases would be prejudicial. But there are other forms of *breast complaints* in which the Waters have been found valuable, particularly in that form described as

SYMPATHETIC CONSUMPTION. *

This form of breast complaint is the result of morbid sympathies extended from some other parts of the body, and more commonly from a diseased stomach or liver. The great *par vagum* nerve, common to both the stomach and lungs, affords a ready medium of sympathy between these two organs. In protracted cases of dyspepsia the stomach often throws out morbid influences to the windpipe and surfaces of the lungs, occasioning *cough, expectoration, pain in the breast*, and many other usual symptoms of genuine consumption. So completely, indeed, does this *translated* affection wear the livery of the genuine disease that it is often mistaken for it. This form of disease comes often under my notice at the Springs, and I frequently witness the happiest results from the employment of the Water in such cases, and the more so because its beneficial effects resolve a painful doubt that often exists in the mind of the patient as to the true character of the disease.

* See " Mineral Waters of the United States and Canadas," by the Author.

BRONCHITIS.

This affection is often met with at the Springs, sometimes as a primary affection of the bronchia, and often as a result of other affections, and especially of derangements of the digestive and assimilative organs. In such *translated* cases we frequently find the *bronchitis* relieved in the same degree that the originally diseased organs are benefited.

CHRONIC DISEASES OF THE SKIN.

The various chronic diseases of the skin are treated with much success by a full course of White Sulphur Waters, in connection with a liberal course of warm or hot sulphur baths. My experience in the treatment of the various forms of *skin disease* with this Water has been large, embracing certainly, in the aggregate, many hundreds of cases; and I would do injustice to such experience, and withhold important hopes from the sufferers from such annoying afflictions, if I failed to express my entire confidence in their employment in such cases. Aided by mild alterative means, proper in themselves, but which of themselves would not generally cure such cases, a thorough course of the Water and baths is entitled to the highest degree of confidence in all such affections.

RHEUMATISM.

The primitive reputation of the Water, and that which at an early day directed public attention to its potency, was derived from its successful employment in rheumatism. The reputation thus early acquired has not been lost, but, on the contrary, established and confirmed by its successful use for three-quarters of a century.

In most rheumatic cases the employment of *warm* or hot *sulphur* baths constitutes a very valuable adjunct in their treatment.

With the Sulphur Water as a drink, and the use of the hot *tub*, *douche* and *sweating* baths of the same water, this place offers the strongest inducements for the resort of persons afflicted with chronic rheumatism that can anywhere be found.

Want of space will not allow me to describe the various forms of rheumatic trouble, and to speak of the relative merits of the waters in the several forms of this disease. I remark, however, that while they prove eminently beneficial in all forms of *chronic rheumatism*, they are more decidedly so in those cases that may properly be termed *muscular*, in distinction from *articular* rheumatism, and this is so whether the cases arise from miasmatic, mercurial, or other more common causes of the disease. *A full course of the waters*, with *baths properly tempered to the demands of the case*, is essential to a perfect cure.

GOUT.

The Gouty are numerous among the *habitues* of the White Sulphur. In proportion as the waters impress the digestive and assimilate organs they benefit *gout.*

Those who come here with confirmed gout often assure me that a course of the waters and baths invariably brings such amelioration of their sufferings for about a year: that is, from one season to another. Hence it is that we so often find the same gouty subject here regularly from year to year.

CHRONIC POISONING FROM LEAD

Is very advantageously treated by a full course of the water and baths. Used with sufficient persistency, they.may well be regarded as the most reliable remedy to which persons thus afflicted can have recourse, and to such I earnestly recommend a trial of them, the more especially because the ordinary remedies in such cases are admittedly very unreliable.

SCROFULA.

Sulphur Waters have long been held in reputation in the treatment of scrofula. Some of the English physicians have thought such waters superior to any other remedy in scrofula. Dr. Salisbury, of Avon, New York, speaks favorably of his experience of their use in such diseases. In the early stages of scrofula the White Sulphur has often been used with decided advantages, but in the advanced stages of this disease I do not consider them at all equal in curative powers to some other mineral waters in this region.

SECONDARY AND TERTIARY SYPHILIS AND MER-CURIO SYPHILIS.

In the unpleasant and dreaded forms of disease sometimes following *Syphilis,* and commonly known as *Secondary or Tertiary Syphilis,* whose symptoms are usually so well marked that they cannot be misunderstood, the *White Sulphur Water,* when carried to its full *alterative* effects, displays its highest curative powers.

After much experience in the use of the Waters in the peculiar forms of disease under consideration, if called upon to name the particular *affection* in which they are most certainly efficacious, I should name *Secondary Syphilis* and its complications with mercurial contamination; because in such cases they exert a specific influence and more certainly than any other remedy, bring relief.

It is proper that I remark, however, that my satisfactory use of the Water in such cases has generally been connected with the moderate use of other means while patients are taking the water, and which, though regarded as remedial will not of themselves

generally eradicate the disease, but when employed in combina-
tion with the Waters, very much hasten the desired result. The
Waters in such cases are the most efficient power; the means employed with them, only valuable adjuvants to hasten their curative
effects.

MERCURIO SYPHILOID.

There is an enfeebled, susceptible and peculiar condition of the
system not unfrequently found to arise as the result of a long-con-
tinued or improper of use mercury in syphilitic disease, and especially in subjects of scrofulous tendency. It seems to be the re-
sultant effect produced by the actions of the two poisons—mercury and the syphilitic virus, costituting a disease *sui generis*, and
neither strictly mecurial or sylphilic, but a hybrid. This peculiar
disease, or state of the system, I designate as *Mercurio Syphiloid*.
I have most frequently met with this peculiar affection in persons
of strong lympathic temperament, and in those of strumous tendency. Such cases exhibit some of the characteristics of ordinary
mercurial disease as well as those of secondary syphilis, but the
disease as a whole is not distinctly marked as either. In such
cases the antidotal effects of the mercury has probably subdued
the virus of the venereal poison, while the joint action of the two
has created a new disease, as loathsome, but not as infectious as
the one for the cure of which the mercury was originally administered. In such cases, the Waters constitute the best remedy
known to me. I know that some may regard my designation of
this *hybrid* disease as singular as its announcement is new, but
nevertheless ample opportunities for many years for examining
such cases establishes, in my judgment, the correctness of the
opinion I express.

EFFECTS OF THE WATER IN INEBRIATION.

During the whole period of my residence at the Springs I have
been interested with the marked power I have seen manifested by
the Waters in *overcoming the desire for the use of ardent spirits* in
those who had been addicted to their imprudent use. I by no
means claim that these Waters should be regarded as a specific
against either the love or the intemperate use of alcoholic drinks,
but simply that a proper use of them is a decided preventative of
that feeling of *necessity* or *desire* for the use of strong drinks which
drives the inebriate to use them, in despite of his own judgment to
the contrary ; or, in other words, that their proper use allays, or
destroys, the aptitude or *nervous craving* for ardent spirits, and to
such an extent that even the habitual drinker and confirmed inebriate feels little or no desire for them while he is properly using
the Waters.

During my long residence at these Springs, I have witnessed hundreds of cases fully justifying the above statement. This peculiar influence of the White Sulphur Water depends, first, upon the action of the *sulphureted hydrogen* gas that abounds in it, and which is an active nervine stimulant, and as such supplies the want the inebriate feels for his accustomed alcoholic stimulant; and, secondly, it depends upon the *alterative* influences exerted by the Waters upon the entire organism. While by its alterative power the entire animal structure is brought into natural and harmonious acting, there is a consequent subsistence of the *cerebral* and *nervous irritation* which always prevails in the habitual drunkard, the abatement of which enables him to exert a moral power greater than he could before, and sufficient to overcome the lessened demand which his old habit, if he retains it in any degree, now makes upon him.

In the initiatory, or forming stage of intemperance, the free use of this Water may be much relied upon to modify, or entirely prevent, the *temptation* for strong drink; and even in the confirmed stage its persevering use may inaugurate a state of the system that will essentially *aid* the sufferer in overcoming the hurtful habit of intemperance. Indeed, if the habitual drinker can be prevailed upon to use the Water properly for some ten days, to the *entire exclusion of alcoholic stimulants*, he will have for the time, at least, but little alcoholic temptation to resist.

Of course, I will not be so misunderstood by any as to suppose that I design even to intimate an opinion that this Water is a *sure* and *permanent cure* for either absolute or threatened inebriation. All I intend to assert in this connection is, *that a proper and continuous use of the Water will very essentially aid the intemperate drinker* to lay aside the inebriating cup and return to soberness.

The *will* of the excessive drinker must necessarily concur to some extent with any effort successfully made for his relief. But while this is so, an auxiliary agent, as innocent in its effects as Sulphur Water, that can so far satisfy the *nervous cravings* of the votary of strong drink as to give him increased power to resist his morbid habit, while at the same time his general health is improved, well deserves, I conceive, the attention of all who need assistance in this direction.

It would be irrational for the inebriate to expect to be cured of his morbid habit by simply visiting the Spring and drinking its Water, however freely, and at the same time (which has been the habit of some) to drink freely also of alcoholic liquors. Such a course could be of no service whatever. Stimulants of whatever kind, in such a case *must be abstained from* while the water is establishing its peculiar action upon the system. This effected, which can ordinarily be accomplished in ten or twelve days, the success of further persistence in the use of the water is *hopeful*, and easily thereafter under the control of the individual who is seeking relief.

The erection of the new Hotel here, adapted, as it will be, for the comfort of *winter boarders*, affords an opportunity to those who may desire to avail themselves of *the aid* of the waters in breaking up established or threatened habits of intemperance, that is well worthy of consideration. The waters are as effectual to that end in cold as in warm weather, while the absence of a large company during the fall, winter, and early spring, is greatly favorable to those periods for such use of them.

USE OF THE WATER BY OPIUM EATERS.

I am occasionally consulted by distant parties who are apprised of the effects of the water in *allaying the desire for ardent spirits*, whether or not it has the same effects in reference to the desire for *opiates*.

Upon this subject I remark that my observations of the influences of the water in *assisting the inebriate* to discontinue the use of alcoholic drinks, when his *will* assents to such discontinuance, very naturally led me to hope that it might afford similar assistance, under a *like consent of the will*, to the *opium eater*. But a good deal of difficulty lies in the way of making reliable observations upon this subject. Opium eaters, even more than excessive drinkers, are indisposed to divulge their morbid propensity to their friends or physician, or to seek, through the aid of either, to be relieved of their hurtful habit; consequently, while personally I have known hundreds of visitants to the Springs who I was satisfied eat opium to excess, and some to very great excess, nevertheless I have had but few cases of inveterate opium eating placed fully under my professional government, with the *single view* of being cured of the habit. Some such cases, however, I have had, in which the sufferers freely and fully communicated to me the fact of their injurious habit, expressed earnest desire to be relieved, and continued during the treatment to exercise all the force of will of which they were capable to render my advice and prescriptions successful. In one of these cases, entirely successful in its treatment, the person had been in the habit for a long time of using not less than *six grains of morphia* daily.

The space allotted to this notice will allow me only now to say that in the few cases alluded to I used the waters very fully, but always *in connection with other means* that I deemed essential—that the success of the combined treatment was very satisfactory—that, in my opinion, the influence of the water, *by lessening the nervous craving for opiates*, materially *aided* in the results, and that such results would not have taken place if the waters had not been used. In the case alluded to, a generous confidence on the part

of the sufferer, which led to prompt observance of professional advice, contributed much, I conceive, especially in the commencement of the treatment, to favorable results.

The most that can confidently be said in favor of the use of the waters in such cases—and all that ought to be said—is that when they are *judiciously used and in connection with proper adjunctive management* and appliances, they essentially *aid* the opium eater in dispensing entirely with the use of that drug. I will only add that, in my management of such cases, I have not found it best to *exclude the entire use of the drug* when the patient *first commences the use of the water*, as I advise shall be done in the case of the inebriate.

I have not hitherto published anything upon this subject, simply from the fact that I am satisfied that the treatment of such cases by the waters to be successful, requires careful professional management, with appropriate adjunctive means,—that the water is only an *efficient aid*, and not a *specific*,—and that the management necessary in connection with it, to give success, depends too much upon the precise circumstances of each case to justify a broad recommendation without numerous and essential qualifications.

DISEASES IN WHICH THE WATERS SHOULD NOT BE USED

I have heretofore mentioned some diseases and states of the system, in which these waters *should not be used*. As mistakes upon this subject are matters of importance, I here recapitulate oft repeated cautions, as to some of the more important diseases, and

First. They should not be used in *Tubercular Consumption.*

Second. They should not be used in *Schirrus* or *Cancer ;* or in that condition of the stomach, or any other organ, threatening to terminate in Schirrhea or Cancer.

Third. They should *never be used* in *Hyperthrophe* or morbid *Enlargement of the heart.* In such cases the use of the water or Baths, always aggravates the disease, and if persisted in, *will very* much *hasten a fatal termination.*

For more than thirty years, by my writings and oral declarations, I have warned the Spring going public against using these waters in enlarged heart ; and yet, sudden deaths from this cause continue occasionally to occur here, either from not knowing, or disregarding such important warnings.

As a Medical Director of these waters, and desirous as I am, that their use shall be strictly confined within their *legitimate power of doing good*, and as a friend to common humanity, I trust that those afflicted with disease that the waters cannot cure, but must aggravate, will be careful to abstain from using them.

The vital importance of these *caveats* to the unfortunate invalids fully justifies the earnestness with which I give them.

CHALYBEATE SPRING.

About forty rods from the White Sulphur is a *Chalybeate Spring*, in which the iron exists in the form of a *carbonate of iron*, the mildest, least offensive, and ordinarily the môst valuable form in which ferruginous waters are found.

For the last twenty years this Water has been considerably used by the class of visitors whose diseases required an *iron tonic*, and its effects have realized the rational hopes that were indulged in it.

BATHS AT THE WHITE SULPHUR.

Warm and *Hot bathing*, especially in highly medicated waters, is a remedy of leading importance in a large number of the cases that resort to mineral waters for relief.

The water used for bathing at the White Sulphur flows from the *Sulphur Springs of which the visitors drink.* When we look at the analysis of this water, and find it to contain about one hundred and fifty *grains of* active *medicinal salts to the gallon*, we cannot fail to see that, so far as the *medication of waters* can favorably affect the *bath* for which they are used, the White Sulphur baths have the strongest claim to confidence, inasmuch as no other waters in America that are used for bathing, except the Washita Springs, in Arkansas, are more highly impregnated with mineral salts.

These baths, in connection with the drinking of the sulphur waters, although not required in every case, are a matter of the utmost importance in a large number of cases in aiding to produce the best effects of the waters.

Impressed with the great value, in fact the absolute necessity to some invalids, of using such baths in connection with the drinking of the water, the proprietors of the Springs have recently greatly enlarged and so remodeled their *bathing establishment* as to make it in every respect satisfactory, it is believed, to those who may desire to avail themselves of its use.

The *bathing house* is large, affording ample accommodations for the bathers. The bathing-rooms are spacious, airy and comfortable, and in addition to the usual *tub baths*, they have erected *douche* baths for the application of streams of *hot* or *warm* water to local parts of the body, and have set apart rooms arranged for receiving *sweating* baths.

The construction of *douche* and *sweating baths* of sulphur water, to be employed under proper circumstances, in connection with the internal use of the water, is a matter of the utmost importance to the successful treatment of numerous cases that resort here for relief.

The new and improved method of heating water for bathing deserves to be especially noted. This is effected by *steam* in the vessel in which it is used, and is a great improvement over the old method of heating mineral waters for bathing. Under the old plan of heating in a boiler and thence conveying the water to the bathing tub, much of its valuable saline matter was precipitated and lost. By this improved method of applying steam to the water in the tub, the heat is never so great in raising the water to the bathing point, as to cause any important precipitation of its salts; hence, they are left in their natural suspension in the waters to exert their specific effect upon the bather. Not only so, by this improved method hot steam may be let into the tub from time to time, as the water cools, so as to keep it essentially of the same temperature during the entire period of bathing, a consideration often of no small importance. This method of heating mineral waters in the tub in which they are used, in connection with the *douche* and *sweating* baths, brings *hot* and *warm bathing* at this place in favorable competition with bathing at naturally hot and warm fountains, and promises to be productive of the same good effects that are experienced from bathing in such fountains.

Persons intending to *bathe in hot sulphur water*, should, previously to doing so, be intelligently instructed under a proper knowledge of their case, as to the precise *temperature* of the bath, and the *length of time* they remain in it. Neglect, or disregard of proper instructions, the relying upon chance or the mere dictum of ignorance upon this subject, has often been the cause, within my knowledge, of aggravation of symptoms, and in several instances, of serious consequences. I state, therefore, for the benefit of bathers in sulphur waters, that such baths, to be *used safely and efficaciously*, must be used with careful reference to their *temperature ;* the *state of the system when employed ;* and the *length of time* the bather remains in them.

SOCIETY AND ITS AMUSEMENTS.

Next to the medical value of the water of the White Sulphur, and the invigorating climate of the place, the company that annually assembles there is most worthy of notice.

The prestige of the White Sulphur for all that is elegant and refined in society is coeval with its early history. For many years it has been the great central point of reunion for the best society of the South, North, East and West, that here mingle together under circumstances well calculated to promote social intercourse, and to call out the kindliest feelings of our nature.

The *cottage system* that has been introduced, although new to American watering places, has proved a complete success, and greatly contributed to the home-like comforts and the socialty of the numerous families assembled here.

Society seems here to meet on common ground, and the different shades of feeling influencing it at home are laid aside, while each individual promotes his own happiness by contributing to the happiness of others.

Here is to be found the statesman who, worn down with labor, and his mind unstrung by the cares of office, seeks from the bracing air, the picturesque scenery, and the genial company, not less than from the health-giving waters, that recuperation of his wasted energies in vain sought for elsewhere.   Here, too, is found the man of letters, seeking rest from thought, and strength for future effort.   The poet, too, is here, to quaff vigor from the sparkling fountain, and new images of beauty from nature's lavish stores that are spread around him ; and here, too, come in crowds those who have ever plumed the poet's fancy to its sublimest flights—beauteous woman—by her presence brightening every prospect and gracing every scene.   Following naturally in her train come those who ever love to bask in beauty's smiles, and find in such scenes the happiest of their youthful hours.   Here, too, congre· gate the reverend clergy, the doctor, the lawyer, the judge, wearied with the burdens of the bench; the man of commerce, the financier, the thrifty planter, the sturdy farmer, and the retired man of wealth and ease.   These, reckoned by thousands, make up the company that annually give tone and character to the White Sulphur, and make it at once the Athens and the Paris of America.

The amusements are various in kind and degree.   No sketch can give more than a faint shadowing of the pleasures of a visit to the Springs.   The freedom from care, the relaxation from bonds which have fettered us to the treadmill of business—the pure mountain air, every breath of which swells the veins and makes the blood tingle with delight—the wild mountain scenery, awakening new thoughts of the grandeur of creation and the mighty power of God—the amenities of social intercourse, relieved from those necessary but vexatious rules of etiquette which hem in fashionable life at home—all these combine to render a visit to the White Sulphur an epoch in life to be looked forward to, and back upon, with pleasurable emotions.

The weary pilgrim, coursing over the burning sands of the East, does not hail the sight of an oasis in mid desert with more joy than the *habitues* of the " White," worn down by cares or trouble, welcome the first glimpse of the sparkling fountain, and the verdant lawns encircled by cottage homes; to him they promise rest, comfort, health, while to others they tell of pleasures past and joys to come.   And why?   For answer, let us briefly sketch the scenes of a single day at the Springs.

The morning has dawned; the forest songster, in saluting the opening day, has softly wakened the sleeper; the full, round face of the sun soon appears above the neighboring mountain peak ; the silvery vapor glides upward from the vale beneath, the fleecy clouds are gone, and the dewy fragrance of the morning air invites to active exercise.   The visitors now gather around the health-giving fountain, and, after quaffing its waters, wend their way to the morning meal. This over, the business of active enjoyment for the day begins.

The pleasant walks that penetrate the lawns and environ the ground invite many to healthful exercise. The billiard saloon, with its numerous tables, entices many votaries; the bowling alleys soon resound with the merry laugh of youth and beauty; and thus the hours glide swiftly away; while from another portion of the grounds is heard the clear, keen report from the pistol gallery, telling how promptly young America is preparing to avenge his insulted honor.

The beautiful rides and drives, with their glorious mountain and intervale scenery, attract some, while the quiet game, the alluring book, or the pleasant companion, solace many others. Thus they take no note of time, save from its loss, until the warning sound of the dinner-bell rings forth the noontide hour, calling to prepare for the mid-day meal. Again the fountain is thronged, and then to the sound of rich-toned music, discoursed by a well-trained band, the crowd, after the hour of preparation has elapsed, assemble in the immense and well-furnished drawing-room for a brief social reunion before partaking of the great meal of the day. Dinner over, the drawing room again becomes the centre of attraction. In this room, during the crowded season, are each day brought pleasantly together a gay and richly dressed assembly. excelled in beauty, manliness and dignity by no other crowd ever assembled within the broad limits of our common country. Here congregate the fairest of the fair from every State, and one can gaze and gaze on beauty until the heart reels in its very fullness.

The company, weary with converse or the promenade, retire to their cottage homes, or to the inviting shade of the wide-spread oaks, underneath which, in by-gone years, the savage danced, or the antlered monarch of the forest tossed his crest, now given up to the happy crowd, who in genial converse wile the hours away until the lengthened shadows and the fragrant air again invite to the walk, the ride, the drive, or other active exercises. Then is heard the summons to a social reunion at tea table, after which the spirit stirring music calls the young and the gay to the giddy whirl of the ball-room. Here pleasure reigns supreme—the heart-toned laugh, the witty word, the amiable repartee, all tell that those assembled here are just sipping the bubbles from the over-flowing cup of joy.

Nowhere else can such a scene be witnessed; nowhere else can such a scene be more innocent than here. Thus flit away the glad hours until the warning night bids to calm repose. Such is, as it were, a shadowy outline of a day at the White Sulphur.

But I cannot close this sketch without mentioning another phase of society at the Springs, and one that must commend itself to every well-ordered mind. I allude to the respectable observ-

ance, by the company generally, of the Sabbath day. Through-
out this entire day a profound quiet pervades the grounds, and the
places of worship are thronged by full and attentive congregations.
Nothing could better evidence the conservative influence of society
here than the respectful and reverential attention with which the
vast concourse honor the sacred claims of the Sabbath.

ANNOUNCEMENT.

GREENBRIER

White Sulphur Springs,

WEST VIRGINIA.

So long and favorably known for their valuable ALTERATIVE WATERS, their charming Summer climate, and the large and fashionable crowds that annually resort to them, will be opened on 1st June.

RATES OF CHARGES FOR SEASON OF 1878, VIZ:

Board, $3.00 per day; $20.00 per week; $75.00 per month, of thirty days.

☞*Special arrangements* may be made for large families that spend the entire season here.

☞*Special Rates* will also be made *for September and October.*

☞*Children* and *Colored Servants* half price.   *White Servants* in proportion to the accommodations furnished.

☞*A First-class Band* will be in attendance to enliven the Lawns and Ball Room.

☞*Masquerade and Fancy Balls* occasionally through the season.

☞*Telegraphic line* in operation to the Springs.

☞A LIVERY is kept for the accommodation of visitors.

☞A well organized LAUNDRY, where all *washing* for the guests will be neatly done at low rates—and to protect ourselves and our guests from loss and outside intrusion, we must insist that the washing of visitors be *confined to our laundry,* for the proper management of which we are always responsible.

☞The Lessees wish it to be distinctly understood that the use of the *White Sulphur Water, Baths* and *Grounds,* will be strictly confined to those *who are the guests of this establishment,* and that their use *will be withheld* from all others, *except upon their paying* $2 per day for such use. Permanent residents of the county *alone* excepted.

PHYSICIAN TO THE SPRINGS.

☞We have the pleasure to inform those who design to visit the Springs that Prof. J. J. MOORMAN, M. D., well known as the author of several valuable books on MINERAL WATERS, and of the work just published on the "MINERAL SPRINGS OF NORTH AMERICA," and for *forty years* the PHYSICIAN TO THE WHITE SULPHUR, will be found at the Springs in that capacity—and that he has associated with him in the practice DR. T. B. FUQUA, formerly of Staunton, Va.

GEO. L PEYTON & CO.

A CARD

WHITE SULPHUR VISITORS.

We are the renters of the White Sulphur Springs *with all their curtaleges*, for which we annually pay a large sum of money, in addition to heavy outlays in preparing the grounds and buildings for the entertainment of a large company, embracing the expenses of the more than 400 employees that are required for the proper management of our establishment, while *our lease covers, and rightfully controls all the sulphur waters of any value in the neighborhood.* The public therefore will readily see that we *cannot*, with any justice to ourselves, allow persons *that are not guests of our Hotel*, to use the waters and the grounds, free from a proper compensation for such use. Hence we ANNOUNCE, and wish it to be distinctly understood, that *all persons boarding or staying* in the neighborhood and using the White Sulphur Waters *while they are not guests* of our establishment, will be charged at the rate of $2 per day for such use, and that this charge will be *invariably enforced*, except in reference to persons permanently residing in the county.

G. L. PEYTON & CO.,

Proprietors White Sulphur Springs.

June 1st, 1878.